সংখ্যাৰ কাহিনী

THE NUMBER STORY

SMALL BOOK ONE

ENGLISH - ASSAMESE

Numbers Teach Children Their Number Names

written and illustrated by

MISS ANNA

Early Reader Edition of *The Number Story 1*
Bronze Medal Winner, 2016 Wishing Shelf Book Award

Library of Congress Control Number: 2018902040

Names: Miss Anna, author.
Title: Number story : numbers teach children their number names / Miss Anna.
Description: Portland, OR: Lumpy Publishing, 2018.
Identifiers: ISBN 978-1-945977-98-5 | LCCN 2018902040
Summary: The pictures and rhymes present stories which introduce numbers 0-10.
Subjects: LCSH Numeration—English--Assamese--Pictorial works--Juvenile literature. | BISAC JUVENILE NONFICTION /
Languages: English--Assamese
Classification: LCC QA141.3 .M57 2018 | DDC 513—dc23

Publisher: Lumpy Publishing
Website: www.missannabooks.com
Email: missanna@missannabooks.com

Paperback: ISBN 978-1-945977-98-5
Printed in the U.S.A.   1 3 5 7 9 10 8 6 4 2

Want to learn our number names?

তুমি সংখ্যাবোৰৰ নামবোৰ জানিব বিচাৰা নে?

It is very easy and a lot of fun!

এইটো বহুত সহজ আৰু আনন্দদায়ক।

Say-along our little jingle

আমাৰ এই সৰু কাহিনী আমি একেলগে গাৰ আহা।

starting from Number One!

আমি সংখ্যা ১ৰ পৰা আৰম্ভ কৰিম।

1

O N E   looks like my one finger.

১ ☆ এক

মোৰ এক আঙুলিৰ দৰে ।

ONE!
এক!

2

TWO   trails a tail.

২ ☆ দুই

এডাল নেজক অনুসৰণ কৰে।

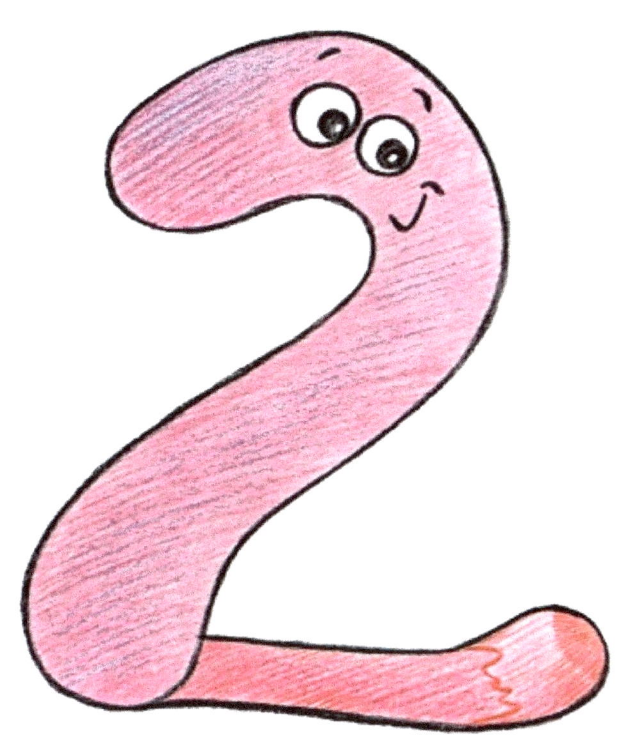

A TAIL! এডাল নেজ!

3

THREE has bumps.

৩ ☆ তিনি

আে ছওখোৰা-মোখোৰা |

ওখোৰা-মোখোৰা!

4

FOUR carries a sail.

৪ ☆ চাৰি

এটা শামুখক কঢ়িয়াই ।

এটা শামুখক!

5

FIVE   is a racing track.

৫ ☆ পাছ

এখন দৌৰৰ ঠাই।

VROOM
VROOM!

6

S I X   curves like a snail.

৬ ☆ ছয়

এটা শামুখ।

A SNAIL! এটা শামুখৰ!

7

BE CAREFUL! IT'S SHARP!

অলপ সাৱধান! এইটো তীক্ষ্ণ!

8

E I G H T   is rollercoaster rails.

৮ ✦ আঠ

ৰালাৰকোষ্টাৰ ৰ পঠ ।

ইপী!
YIPPEE!

9

NINE   is a bubble on a stick.

৯ ✩ ন

লাঠি এডালত পানীৰ বুৰবুৰণী ।

A BUBBLE!

পানীৰ বুৰবুৰনী!

10

TEN is an eye of a whale.

১০ ☆ দহ

তিমি মাছৰ এটা চকু।

HELLO!
নমস্কাৰ!

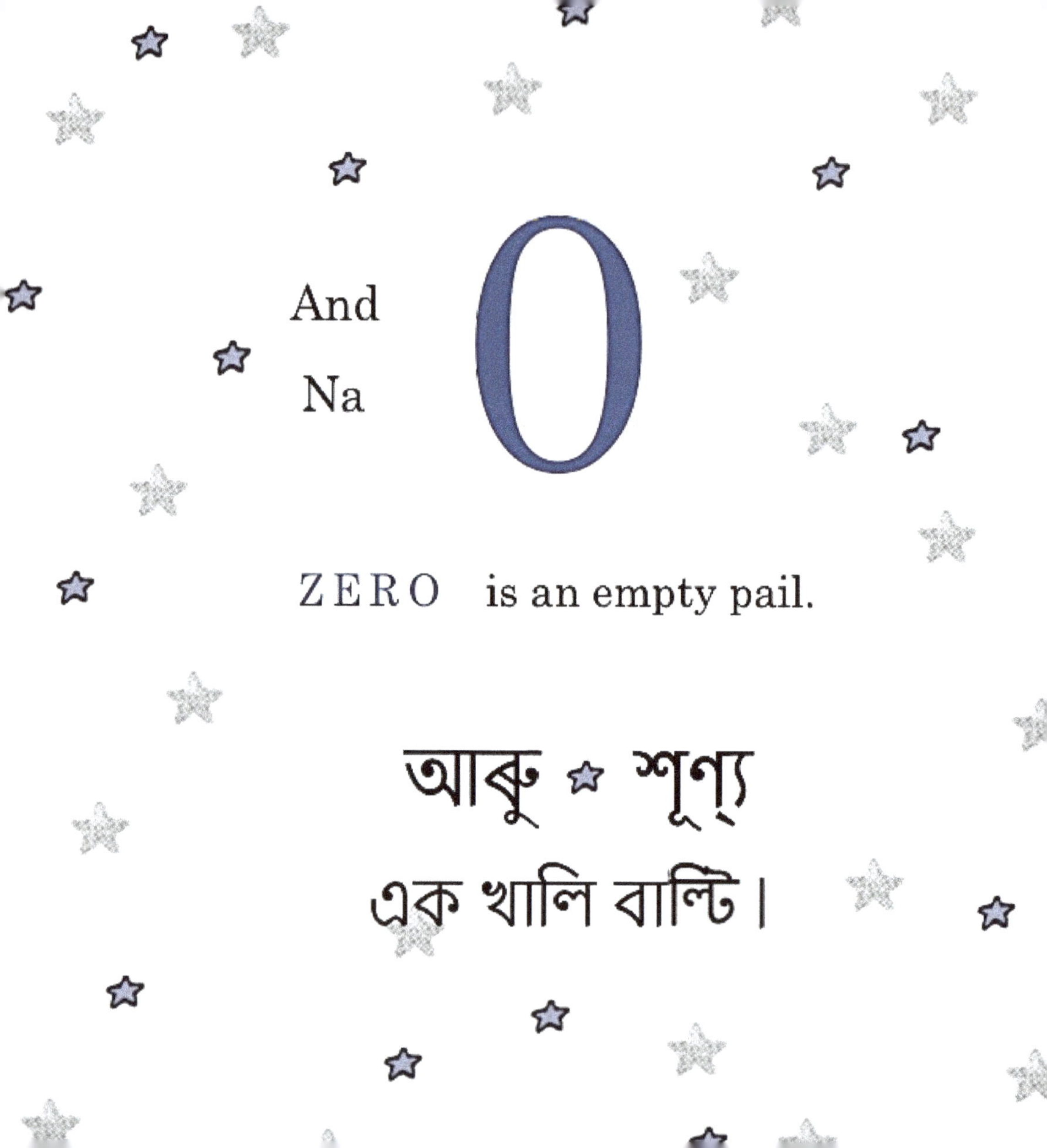

And
Na
0
ZERO   is an empty pail.
আৰু ✦ শূন্য
এক খালি বাল্টি।

IT'S EMPTY!
এইটো খালি!

Thank you for playing with us today.

We had a lot of fun too!

আজি আমাৰ লগত খেলাৰ বাবে ধন্যবাদ।
আমিও তোমাৰ লগত খেলি
বহুত আনন্দ লাভ কৰিলো।

We are your Number friends,
Zero to Ten,
Who will be here for you~

আমি সংখ্যাবোৰ তোমাৰ বন্ধু
শূণ্যৰ পৰা দহ লৈ।
আমি তোমালোকৰ বাবে সদায় ইয়াত আছো!

Bye-bye now!
See you again soon!

এতিয়া বাই-বাই!
আকৌ লগ পাম অতি সোনকালে!

The Numbers are *SINGING* too!

To sing-a-long, look for Miss Anna Number Story
at your favorite music store like iTUNES.

MP3

Numbers 0-10
IDENTIFYING
& COUNTING

Numbers 11-20
& Ordinals
first, second, third...

Numbers 0-100
& Place Values
ones, tens, hundreds...

About Clocks
& Telling Time
hours, minutes, seconds

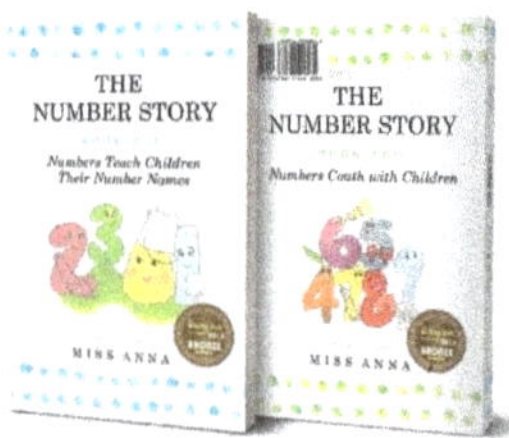

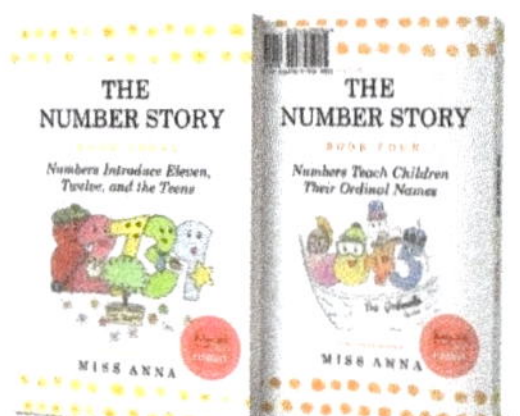

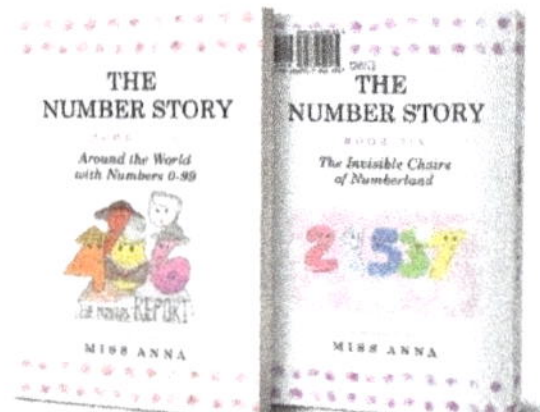

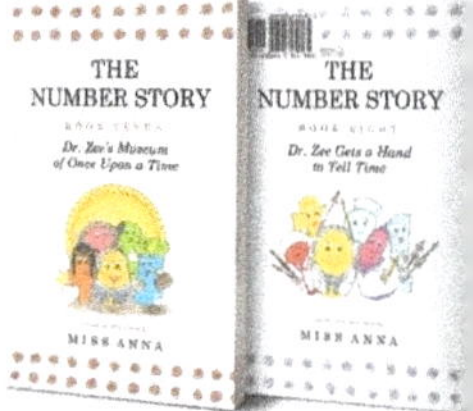

Number Story 1 & 2

isbn: 978-0-996216-48-7

Number Story 3 & 4

isbn: 978-1-945977-01-5

Number Story 5 & 6

isbn: 978-1-945977-06-0

Number Story 7 & 8

isbn: 978-1-949320-40-4

For more Miss Anna books to love,
visit us at

w w w . m i s s a n n a b o o k s . c o m

Numbers are working hard all over the world!
Come Travel the World with Us!